YOUR KNOWLEDGE HAS VALUE

- We will publish your bachelor's and master's thesis, essays and papers

- Your own eBook and book - sold worldwide in all relevant shops

- Earn money with each sale

Upload your text at www.GRIN.com
and publish for free

Bibliographic information published by the German National Library:

The German National Library lists this publication in the National Bibliography; detailed bibliographic data are available on the Internet at http://dnb.dnb.de .

Imprint:

Copyright © 2016 GRIN Verlag
Print and binding: Books on Demand GmbH, Norderstedt Germany
ISBN: 9783668659476

This book at GRIN:

https://www.grin.com/document/383491

Mike Nkongolo, Nicholas Scott Telford

Investigating relationships between student marks and majors taken. A descriptive and inferential statistics using SAS

GRIN Verlag

GRIN - Your knowledge has value

Since its foundation in 1998, GRIN has specialized in publishing academic texts by students, college teachers and other academics as e-book and printed book. The website www.grin.com is an ideal platform for presenting term papers, final papers, scientific essays, dissertations and specialist books.

Visit us on the internet:

http://www.grin.com/

http://www.facebook.com/grincom

http://www.twitter.com/grin_com

Report: Investigating Relationships Between Student Marks and Majors Taken

Nicholas Scott Telford
Mike Nkongolo

June 16, 2016

Contents

Aims

This project seeks to discover any correlation between students who did Mathematics and Applied Mathematics and their chance of passing. This investigation will try to determine if for the double major students, they are more likely to pass without repeating a single time than the students only taking one major.

Chapter 1

Problem Definition

1.1 Introduction

This project plan was created to detail the sections of the project and the specific steps that we planned to carry out during the course of the investigation. It helped to ensure the timeous completion of the project, as well as to meet all requirements and goals. Our project plan details the following:

- Outline of the environment of the business the project is being done for
- Outline of the deliverables that are needed from our project
- Investigation into risks and how to deal with them during the course of the project

In addition, we hope to cover the following points:

- The importance of the work
- The specific steps we took to arrive at the end goals
- The questions that are answered by carrying out the data transformation, and subsequently, the statistical analysis

From examining the business requirements, as well as the specific project goals, we determined the direction to take to start with. We used the plan to understand the particular business environment (in this case, a University) that the problem is related to and thus were able to take steps specific to that environment. This allowed us to understand which statistical methods were the most appropriate and thus we performed the appropriate tests and procedures to answer our questions.

1.2 Problem Visualisation

The dataset we are working with in this investigation is related to the personal information of students at the University of the Witwatersrand. This includes their details such as age and sex. In addition, the subject choices of students are given as well as their final marks and grades for each of their subjects, in each of their years of study. We are looking to determine whether there is a relationship between students who either majored in one subject (Applied Mathematics only) or in two subjects (both Mathematics and Applied Mathematics) and their chance of passing all 3 years without repeating. We will also investigate any other possible meaningful trends within the data.

The work being done is important because it will allow the University of the Witwatersrand to respond to common trends observed in the data. For example, if it is observed that a certain group of students performed worse than another, it will enable the faculty within the University to put measures in place to help increase the pass rate of the struggling students. In order to achieve this, we must follow the project plan detailed in the next section.

1.3 Plan Overview

In this investigation, we are required to:

- Clean the dataset, removing any erroneous records and other sources of problematic data that may skew our results,
- Transform the dataset to conform to the needs of a statistical analysis, removing and modifying variables where necessary,
- Apply statistical methods over the dataset using the software SAS in order to extract important information from it,
- Create summary and contingency tables to better understand the data,
- Detect patterns by plotting graphs to gain useful knowledge from the data,
- Use the information gained to answer questions raised in our project plan.

1.4 Risk Management

A project of this nature does incur significant risk. For example, the dataset we are working with may have an overwhelming amount of erroneous data. This would skew our results heavily and the outcome of the project would be unsuccessful. In addition, we must consider that the sample size of the data we are working with may turn out to be too small to gain any meaningful information from it.

Therefore, we must take steps to remove as much erroneous data as possible. Processing the data before running statistical analyses on it will prevent problems from occurring later on and having to backtrack. If there is simply too much erroneous data or if the dataset is too small, then there is not a simple solution to achieve meaningful results. The investigation may need to be repeated in the future, by taking a larger dataset from the source.

Chapter 2

Data Collection and Preparation

The data used for this investigation was not initially clean enough for meaningful analysis to be performed on it. We thus outline the various steps taken to prepare this data in order to make it ready for statistical tests.

2.1 Source of the Data

The educational dataset used for this project have been provided by the System Support Consultant from the Academic Information System Unit at the University of the Witwatersrand. The dataset contains students who were registered for Applied Mathematics or Mathematics between 2010 and 2015 and enrolled for a BSc degree of 3 years.

2.2 Data Dictionary

The dataset we utilised for the purposes of this project contained a number of varied attributes. These attributes had different data types, and for the purpose of the analysis, different importance. The dataset is described in the data dictionary in Table 2.1 below:

Table 2.1: Data dictionary

Field Name	Data Type	Constraints	Description
Calendar Instance Year	Number	Between 2010 to 2015	Year in which student studied
Area	Char	A-Z	Student's main study field
Program Code	Char	A-Z, 0-9	Code for their overall degree
Program Title	Char	A-Z, 0-9	Name of their degree
Year of Study	Char	YOS1, YOS2, YOS3	Current year of the degree
Unit Code	Char	A-Z, 0-9	Identification code of course
Unit Title	Char	A-Z, 0-9	Name of course
Unit Attempt Status	Char	Enrolled or not	Whether student is enrolled
Index Key	Integer	Unique for each student	Unique key to identify a student
Age	Integer	0 and up	Age of student
Sex	Dichotomous	M or F	Gender of student
Final Mark	Number	0 to 100	Final mark obtained in course
Final Grade	Char	Result code	Whether the student passed

2.3 Issues with the Data

After completing a thorough examination of the dataset, a number of key issues were noted. The dataset, as is commonly the case, was not very clean and contained a number of issues that needed to be sorted out. The first thing to be noticed was that there were a number of observations with missing values in the "Final Mark" of the students. These missing or incomplete observations have the potential to skew the results of the statistical analysis if they were taken into account. Thus, they must be dealt with.

In addition, some final results of students were erroneously recorded. While the range of marks can only be between 0 and 100, some observations contained values outside this range, which were clearly incorrect. These incorrect marks likely came about as a result of human error whilst recording students' achievements.

Next, some attributes in the dataset were insignificant. Attributes such as "Unit Attempt Status" and "Program Title" did not provide any new information and thus were redundant.

Finally, there were an abnormally large number of students that had '0' recorded as their final mark, which was likely to skew the results of the analysis unfairly. We looked at factors that caused these types of results to occur.

4

2.4 Data Cleaning

Referring to the previous section, all the issues we outlined were tackled as follows in this section. Firstly, all data with missing observations was removed from the dataset. This eliminates any potential problems that these missing values could cause, such as not allowing sums or averages to be calculated. The data analysis will be focused strictly on only those observations that are complete.

Next, all observations that had a final mark outside the range of 0-100 were removed. As it is impossible to tell what mark these students actually achieved, assumptions cannot be made and it is more beneficial to just remove them completely from the dataset. Inserting any assumed values would just skew the dataset in an unnatural way.

The attributes "Calendar Instance Year", "Program Code", "Program Title", "Unit Title", and "Unit Attempt Status" were then removed from the dataset. All necessary information required to match students to a particular course can be found from the "Program Code" attribute. This one attribute tells us whether the course was a Mathematics course (indicated by containing "MATH") or whether it was an Applied Mathematics course (indicated by "APPM"). In addition, it also tells us which year or level each course was at, indicated by the numbers following the aforementioned prefixes. The "Year of Study" attribute tells us how many years the student had completed and thus if they managed to complete the required number of years for a major.

In terms of the large numbers of 0s in the data for final marks, we noticed that these were often associated with Final Grades such as "FAB" or "FNQ". FAB refers to failing due to absence [Wits 2016], thus students were given a 0 simply for not attending the final exam, and the 0 may not truly reflect their real result. FNQ stood for students that did not qualify to write the exam [Wits 2016], through not submitting assignments and writing tests. These students were also automatically given 0 as their final mark. Therefore, we removed all observations with these Final Grades to ensure our data was not skewed towards 0 too greatly.

2.5 Data Transformation

With the remaining useful attributes we selected, it was then necessary to transform the data further to prepare it for analysis. We sorted the data according to the Index Key attribute, meaning that each individual student would have all their observations and courses taken put together. The specifics of this investigation were as follows - we wanted to determine if there was a relationship between students that majored (did all 3 years) in either Applied Mathematics only or both Mathematics and Applied Mathematics and whether they passed or had to repeat at least 1 year. Since the data was taken from the year 2010 to 2015, there is no data about students that took courses in years prior to 2010, and who are finishing their majors in 2010 or 2011. In addition, some students might have only started studying in 2014 or 2015, meaning they would not have finished their majors yet and there would thus be insufficient data to determine whether they have passed or repeated. Finally, some students may have simply taken one year of mathematics courses as a requirement of another degree and not done any more mathematics courses since.

As a result of this, we removed the aforementioned data as it did not contribute to the analysis we intended to perform. This involved creating a new table and only inserting those students (marked by their Index Keys) that took courses in YOS (year of study) 1, 2 and 3. Next, it was necessary to separate the data further into different tables according to students that majored in both Mathematics and Applied Mathematics and students that only majored in Applied Mathematics. A student that majored in a field was defined as one that did all 3 years of courses in that field. This allows for analysis to be performed on these two different groups, in order to see how their chances of passing match up.

Next, in order to determine the relationships between students passing or repeating and their course choices, we grouped the data according to each index key (there is one key per student) and placed it in a summary table. This meant that we needed to calculate the average mark for each student for all courses they took. Any students that repeated any course at least once were recorded as 'Repeat' students in the Final Grade column. Students that didn't repeat a single time were recorded as 'Pass' students. We also kept a column indicating whether each student was single major or double major.

In addition to these tables, we also further broke down the tables with data split according to double major students that passed, double major students that repeated, single major students that passed, and single major students that repeated. This was done to perform descriptive statistical analysis on these sets of data. In addition, we opted to look at the relation between age and pass chance, sex and pass chance and field of study and pass chance, separating data into further tables in order to achieve this (for example, putting all Engineering students in one separate table, and putting male and female students in different tables). For the age analysis, age values were binned according to whether they were under 25 or 25 and over. This allows for the creation of contingency tables during chi-square analysis - the counts of the number of students that passed and failed according to whether they were male or female can be retrieved, for example.

Finally, we also created summary tables to perform t-tests on with two columns, one representing whether each student was a single major student or a double major student (marked as 'A' or 'MA') and the other showing what their final marks were. This allows for averages and standard deviations of single major marks and double major marks to be calculated, respectively, and used in a t-test. To examine age, gender and field using t-tests, we also created tables showing, for example, whether each student was male or female and their corresponding final marks, and similar tables for the other two tests.

Since the student marks already exhibited a roughly normal distribution, it was not necessary to perform any further normalisation steps on the dataset.

6

Chapter 3

Statistical Analysis

The overall shape of the data was first examined with a frequency distribution, as seen below:

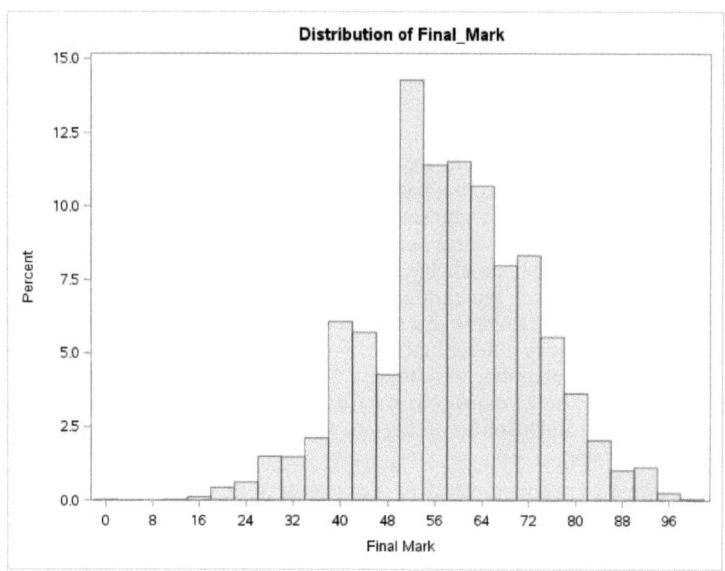

Figure 3.1: Frequency distribution of final marks for all students

As we can see in Figure 3.1, the data follows a relatively normal distribution in terms of final marks for students.

3.1 Descriptive Statistics

In this section, we examined the central tendency, variation and shape of distribution of the data. We utilised built-in functions in SAS [SAS 2010a] in order to perform the calculations (Proc Means), as seen in the code in the appendix.

3.1.1 Central Tendency

The formula used by SAS to calculate the mean in the following section was as follows (from class notes):

$$Mean : \bar{x} = \frac{\sum_{i=1}^{n} x_i}{n} \tag{3.1}$$

The central tendency of the unsplit dataset (containing all major students) was as follows:

Table 3.1: Central Tendency Statistics for all Major Students

	Mean	Median	Mode	No.
Overall	58.224	58	50	4967
Math + Applied Math Major	57.667	57	50	2995
Applied Math Major Only	59.071	60	50	1972

As we can see in Table 3.1 above, the average value of the final mark for all major students is 57.542. This means that, on average, a student was more likely to pass overall than have to repeat. The middle observation, or median, was 58. The most often occurring mark was 50. We also examined the values for the mean, median and mode of the tables after being split from the original data. For students that majored in both mathematics and applied mathematics, the average final mark across all subjects was 57.667, with a median of 57 and mode of 50. In contrast, for students who majored only in applied mathematics, the average final mark was found to be 59.071, which is slightly higher than that of the double major students.

We then looked at a further broken-down set of tables, broken into students that passed and students that repeated, for both single major and double major students. As described in the previous section, we grouped data according to index key, meaning averages of each student's marks for all their subjects were taken. Thus, statistical measures were taken on these averages rather than the full dataset in the following segment. The central tendency statistics for these tables are displayed below:

Table 3.2: Central Tendency Statistics Split by Passes/Fails

	Mean	Median	Mode	N
Math + Applied First-Time Passes	65.634	63.462	57.706	6
Math + Applied Repeats	54.684	54.188	49.708	11
Applied Only First-Time Passes	66.579	64.091	59.000	12
Applied Only Repeats	55.165	54.531	61.000	9

In Table 3.2 above, we see the mean, median and mode calculated for each portion of the dataset. An interesting point to note is the percentage of students that passed and failed for the sets of students that did a single major and those that did a double major. For double major students, only 35.465% of students passed their courses on the first try, while 64.535% of students repeated a subject at least once. In contrast, 58.371% of single students passed their courses in the first try, while 41.629% repeated at least once. Therefore single major students had a higher pass rate.

Table 3.3: Central Tendency Statistics Split According to Faculty, Sex, Age

	Mean	Median	Mode	No.
Science Students	58.873	57.409	56.833	278
Commerce Students	58.147	56.833	49.667	57
Engineering Students	66.993	65.208	61.000	86
Male Students	60.589	58.727	56.833	255
Female Students	60.053	59.448	61.000	138
Under 25 Students	60.819	59.500	60.000	285
25 and Over Students	59.328	56.941	61.000	107

The mean final marks for engineering students and students under 25 years old were higher than their counterparts, as seen in Table 3.3. Other averages were relatively similar to each other.

3.1.2 Variation

The formulae used by SAS to calculate the variance and standard deviation in the following section were as follows (from class notes):

$$Variance : s^2 = \frac{\sum_{i=1}^{n}(x_i - \bar{x})^2}{n - 1} \tag{3.2}$$

$$StandardDeviation : s = \sqrt{s^2} \qquad (3.3)$$

Table 3.4: Variation Statistics for all Major Students

	Min	Max	Range	Q1	Q3	Variance	Std Deviation
Overall	0	100	100	50	67	192.249	13.865
Math + Applied Math Major	13	100	87	50	67	196.236	14.008
Applied Math Major Only	0	97	97	50	68	185.103	13.605

As seen in Table 3.4, variances for all 3 groups were relatively high, indicating the final marks of students varied an average of around 14 points from each other.

We then grouped students together by their index keys, causing the marks for each of their subjects to need to be averaged.

Table 3.5: Variation Statistics for Different Major Combinations

	Min	Max	Range	Q1	Q3	Variance	Std Deviation
Math + Applied First Time Passes	52.764	88.059	35.294	59.810	69.471	63.745	7.984
Math + Applied Repeats	42.833	76.444	33.611	50.952	57.333	30.215	5.497
Applied Only First Time Passes	52.000	92.333	40.333	59.545	72.667	89.930	9.483
Applied Only Repeats	44.368	69.667	25.298	51.415	59.481	23.654	4.864

The result is smaller variances and ranges, as seen in Table 3.5. The average marks of first-time passing students generally had a greater variance.

Table 3.6: Variation Statistics for Different Faculties, Ages and Genders

	Min	Max	Range	Q1	Q3	Variance	Std Deviation
Science Students	43.107	88.059	44.952	53.545	62.615	65.532	8.095
Commerce Students	42.833	82.833	40.000	52.423	63.059	71.337	8.446
Engineering Students	51.250	92.333	41.083	59.750	73.667	108.386	10.411
Male Students	42.833	91.333	48.500	53.600	66.412	98.770	9.938
Female Students	43.107	92.333	49.226	54.800	63.059	61.628	7.850
Students Below 25	43.107	92.333	49.226	54.462	64.647	85.319	9.237
Students 25 and Older	42.833	90.333	47.500	53.000	63.333	86.158	9.282

10

Variances and standard deviations were evenly spread when looking at different faculties, ages and genders and their relationships with marks, as seen in Table 3.6.

3.1.3 Shape of the Distribution

The formulae used by SAS to calculate the skewness and kurtosis in the following section were as follows (from class notes):

$$Skewness = \frac{\sum_{i=1}^{n}(x_i - \bar{x})^3}{(n-1)s^3} \tag{3.4}$$

$$Kurtosis = \frac{\sum_{i=1}^{n}(x_i - \bar{x})^4}{(n-1)s^4} \tag{3.5}$$

Table 3.7: Shape Statistics for All Majors

	Skewness	Kurtosis
Overall	-0.081	0.011
Math + Applied Math Major	-0.072	0.028
Applied Math Major Only	-0.086	-0.019

For all 3 distributions in Table 3.7, the data was skewed slightly to the right (as given by the skewness values being negative) and had a relatively flat peak (kurtosis values being close to 0).

Table 3.8: Shape Statistics for Different Major Combinations

	Skewness	Kurtosis
Math + Applied First Time Passes	0.987	0.631
Math + Applied Repeats	0.682	1.838
Applied Only First Time Passes	0.859	0.067
Applied Only Repeats	0.345	-0.016

For the double major students, for both first time passers and repeaters, the data was skewed heavily to the left (positive skewness values), with repeaters having a more pronounced peak than the first time passers (higher kurtosis value). Single major applied student data (first time passers) was also skewed heavily to the left but had a much flatter peak, while single major repeaters had a more symmetrical distribution and a flat peak.

11

Table 3.9: Shape Statistics for Different Faculties, Ages and Genders

	Skewness	Kurtosis
Science Students	1.075	1.407
Commerce Students	0.972	1.277
Engineering Students	0.704	-0.185
Male Students	0.915	0.473
Female Students	1.428	3.434
Students Below 25	1.034	1.127
Students 25 and Older	1.135	1.343

All sets of data in Table 3.9 displayed heavy skewing to the left, as a result of positive skewness values. Peaks were mostly pronounced too, especially for female students, due to the high skewness values. Only engineering student data had a particularly flat peak.

3.2 Inferential Statistics

Once again, SAS functions (Proc Ttest [SAS 2010c] and Proc Freq with chisq option [SAS 2010b]) were used to do the calculations for our data.

3.2.1 Two-sample t-tests

The formula used by SAS to calculate the T-statistic in the following section was as follows (from class notes):

$$T = \frac{\bar{x}_1 - \bar{x}_2}{s_p\sqrt{\frac{1}{n_1} + \frac{1}{n_2}}} \tag{3.6}$$

We performed two-sample t-tests on the data to determine whether there was a difference in the average final marks for students in different situations. The general hypotheses were as follows:

- H_0: There is no significant difference between the student mark group means.
- H_a: There is a significant difference between the student mark group means.

We started off by comparing the means of final marks between students that did a single major and students that did a double major. We took random samples, approximately equal in size, from each group of data. The results are seen in Figure 3.2 below:

Major	Method	Mean	95% CL Mean		Std Dev	95% CL Std Dev	
A		56.2037	55.0096	57.3978	13.3972	12.6046	14.2970
MA		57.4179	56.2192	58.6166	13.3796	12.5842	14.2832
Diff (1-2)	Pooled	-1.2142	-2.9040	0.4757	13.3885	12.8169	14.0138
Diff (1-2)	Satterthwaite	-1.2142	-2.9040	0.4757			

Method	Variances	DF	t Value	Pr > |t|
Pooled	Equal	965	-1.41	0.1589
Satterthwaite	Unequal	964.92	-1.41	0.1589

Figure 3.2: Results of t-test for comparing single major and double major student marks

We selected the 'pooled' method of t-value calculation, since the group variances are approximately equal. We discovered a t-value of -1.41. At $\alpha = 0.05$, the critical values for a two-tail test are -1.96 and +1.96. Since our calculated value falls between these two values, we fail to reject the null hypothesis. There is no significant difference between the average student marks between single major (indicated as 'A' in the table) and double major students (indicated as 'MA'). However, even though the means are not different, this does not necessarily mean that the amounts of passes and repeats will be the same too, as we will see later.

We then looked to see if the mean marks between students in the science faculty and students from other faculties were different.

Area	Method	Mean	95% CL Mean		Std Dev	95% CL Std Dev	
O		61.4182	59.3151	63.5214	10.4902	9.1990	12.2064
S		57.1519	55.6309	58.6730	7.6262	6.6918	8.8665
Diff (1-2)	Pooled	4.2663	1.6911	6.8415	9.1635	8.3372	10.1729
Diff (1-2)	Satterthwaite	4.2663	1.6854	6.8472			

Method	Variances	DF	t Value	Pr > |t|
Pooled	Equal	195	3.27	0.0013
Satterthwaite	Unequal	177.07	3.26	0.0013

Equality of Variances				
Method	Num DF	Den DF	F Value	Pr > F
Folded F	97	98	1.89	0.0018

Figure 3.3: Results of t-test for comparing marks between different faculties

As we can see, the discovered t-value is 3.27. This value falls outside the range of -1.96 to +1.96. As a result of this, we must reject the null hypothesis. This means that there is a relationship between the field or area the student is in and their average final mark. Students in other fields (indicated as 'O' in the table) displayed a higher average mark compared to students in the science field (indicated as 'S').

Next, we looked to see if the mean marks between students of different genders were different.

Sex	Method	Mean	95% CL Mean		Std Dev	95% CL Std Dev	
F		59.8996	57.9762	61.8229	8.3018	7.1462	9.9065
M		57.5928	55.6837	59.5019	8.2403	7.0933	9.8332
Diff (1-2)	Pooled	2.3068	-0.3806	4.9941	8.2711	7.4214	9.3422
Diff (1-2)	Satterthwaite	2.3068	-0.3806	4.9941			

Method	Variances	DF	t Value	Pr > |t|
Pooled	Equal	146	1.70	0.0919
Satterthwaite	Unequal	145.99	1.70	0.0919

Figure 3.4: Results of t-test for comparing marks between different genders

We found a t-value of 1.70. This falls inside the required range, leading us to fail to reject the null hypothesis. There is no significant difference between the average marks of students with different genders, and as such, gender has no meaningful effect on performance.

Lastly, we looked to see if the mean marks between students of different ages were different.

Age	Method	Mean	95% CL Mean		Std Dev	95% CL Std Dev	
O		59.0609	57.2596	60.8622	9.0315	7.9248	10.5002
U		58.7099	56.9124	60.5073	9.0122	7.9079	10.4778
Diff (1-2)	Pooled	0.3511	-2.1778	2.8800	9.0218	8.2103	10.0128
Diff (1-2)	Satterthwaite	0.3511	-2.1778	2.8800			

Method	Variances	DF	t Value	Pr > |t|
Pooled	Equal	196	0.27	0.7845
Satterthwaite	Unequal	196	0.27	0.7845

Figure 3.5: Results of t-test for comparing marks between different ages

We found a t-value of 0.27. This falls within the required critical value range, meaning we fail to reject the null hypothesis. Students below 25 years of age (indicated by 'U' in the table) did not have a significantly higher or lower average final mark compared to students 25 years and older (indicated by 'O').

14

3.2.2 Chi-Square Tests

The formula used by SAS to calculate the chi-square value in the following section was as follows (from class notes):

$$\chi^2 = \sum_{i=1}^{k} \frac{(O_i - E_i)^2}{E_i} \tag{3.7}$$

We first performed a chi square test to examine whether there was a relationship between the combination of majors a student took and their final result. The hypotheses we intended to test were as follows:

- H_0: There is no relationship between the majors a student takes and their final result.
- H_a: There is a relationship between the majors a student takes and their final result.

We created a contingency table from the data as seen in Table 3.10 below. We then ran the SAS chi-square test function on this contingency table. The results were as follows:

Table 3.10: Contingency Table for Math + Applied Math vs Applied Math Only

	Pass	Repeat	Total
Math + Applied Math Students	61	111	172
Applied Math Only Students	129	92	221
Total	190	203	393

Degrees of freedom: 1 ; Chi-squared value: 20.322

As we can see, the chi-square value was calculated to be 20.322 at 1 degree of freedom. At a degree of freedom 1 and $\alpha = 0.05$, the critical value of chi-square is 3.841. Since the calculated value is greater than the critical value, we must reject the null hypothesis. This means that there is indeed a relationship between the majors a student takes and their result. When students only major in applied mathematics, they are significantly more likely to pass than if they are doing a double major. This difference is not due to chance.

We then wanted to investigate whether gender had any effect on chances of passing. Our hypotheses were as follows:

- H_0: There is no relationship between the gender of a student and their final result.
- H_a: There is a relationship between the gender of a student and their final result.

We created another contingency table from the data as seen in Table 3.11 below. We then ran the SAS chi-square test function on this contingency table once more. The results were as follows:

Table 3.11: Contingency Table for Male vs Female

	Pass	Repeat	Total
Male Students	125	130	255
Female Students	65	73	138
Total	190	203	393

Degrees of freedom: 1 ; Chi-squared value: 0.132

The chi-square value was calculated to be 0.132 at 1 degree of freedom. At a degree of freedom 1 and $\alpha = 0.05$, the critical value of chi-square is 3.841. Since the calculated value is smaller than the critical value, we fail to reject the null hypothesis. This means that there is no relationship between the gender of a student and their result. Students are statistically equally likely to pass or repeat whether they are male or female. Any differences are due to chance.

We then examined whether there was a relationship between the field the student is studying in and their result. We formulated the following hypotheses:

- H_0: There is no relationship between the field of study of a student and their final result.
- H_a: There is a relationship between the field of study of a student and their final result.

We created a third contingency table from the data as seen in Table 3.12 below. We then ran the SAS chi-square test function on this contingency table again. The results were as follows:

Table 3.12: Contingency Table for Different Fields

	Pass	Repeat	Total
Commerce	19	26	45
Engineering	63	21	84
Science	108	156	264
Total	190	203	393

Degrees of freedom: 2 ; Chi-squared value: 30.419

The chi-square value was calculated to be 30.419 at 2 degrees of freedom. At a degree of freedom 2 and $\alpha = 0.05$, the critical value of chi-square is 5.991. Since the calculated value is greater than the critical value, we must reject the null hypothesis. This means that there is a relationship between the faculty or field the student is studying in and their result. Students in the engineering field were shown to be statistically more likely to pass without repeating than those in the commerce and science fields. This difference is not due to chance.

We then examined whether there was a relationship between the age of a student and their result. We formulated the following hypotheses:

- H_0: There is no relationship between the age of a student and their final result.
- H_a: There is a relationship between the age of a student and their final result.

We created a final contingency table from the data as seen in Table 3.13 below. We then ran the SAS chi-square test function on this contingency table again. The results were as follows:

Table 3.13: Contingency Table for Different Ages of Students

	Pass	Repeat	Total
Under 25	143	142	285
25 and Older	47	60	107
Total	190	202	392

Degrees of freedom: 1 ; Chi-squared value: 1.217

The chi-square value was calculated to be 1.217 at 1 degree of freedom. At a degree of freedom 1 and $\alpha = 0.05$, the critical value of chi-square is 3.841. Since the calculated value is less than the critical value, we fail to reject the null hypothesis. This means that there is no signficant relationship between how old the student is and their result.

Chapter 4

Delivery of Results

From our results, we determined that students taking only one major were more likely to perform better than those doing a double major. This may be due to the increased workload, which would not allow students to devote an appropriate amount of time to each subject. For this reason, it may be worthwhile to look at avenues that may help improve the chance of a double major student's success. Extra tutorials, for example, could be introduced to assist any students who feel they are struggling and likely to repeat a course. Another possible option would be to reduce the number of assignments given out in each course, allowing students enough time to focus on everything.

We also noticed that students from different faculties had different chances of passing. This may be due to students taking on other field-specific subjects and courses in addition to their mathematics courses, increasing their workload and giving them less time to prepare for everything. Students may need to be warned before they commence their studies that the amount of subjects they are taking on is too great, based on past students' results.

4.1 Conclusion

Our investigation set out to determine if there was a relationship between the number of majors a student took and their likely achievement. We cleaned the data, removing missing values and unnecessary attributes. We then transformed the data and performed statistical tests to acquire both descriptive and inferential statistics. Our investigation discovered the association between majors taken and results obtained. Students majoring only in applied maths were found to be more likely to pass than students doing a double major. In addition, students in different fields had different chances of succeeding. The University can now determine appropriate actions to take to increase its students' performance.

Bibliography

[SAS 2010a] SAS. *Base SAS(R) 9.2 Procedures Guide: Statistical Procedures, Third Edition*, 2010. Retrieved May 2016, from http://support.sas.com/documentation/cdl/en/proc/61895/HTML/default/viewer.htm#a000146729.htm

[SAS 2010b] SAS. *Base SAS(R) 9.2 Procedures Guide: Statistical Procedures, Third Edition*, 2010. Retrieved May 2016, from http://support.sas.com/documentation/cdl/en/procstat/63104/HTML/default/viewer.htm#procstat_freq_sect027.htm

[SAS 2010c] SAS. *SAS/STAT(R) 9.2 User's Guide, Second Edition*, 2010. Retrieved May 2016, from https://support.sas.com/documentation/cdl/en/statug/63033/HTML/default/viewer.htm#ttest_toc.htm

[Wits 2016] Wits. University of the witwatersrand - explanation of result codes. *Wits Examinations and Graduations Office*, 2016.

Chapter 5

Appendix - Code Used

```
/* Nicholas Telford (671271) and Mike Nkongolo (1171635)
SAS code for Data Analysis Project
Submitted: 16 June 2016 */

FILENAME REFFILE '/folders/myfolders/Project_Data_2016.xls ';

PROC IMPORT DATAFILE=REFFILE
            DBMS=XLS
            OUT=WORK.PROJDATA;
            GETNAMES=YES;
RUN;

PROC CONTENTS DATA=WORK.PROJDATA; RUN;


/* Clean the dataset */
Data PData (Drop=Program_Code Drop= Program_Title Drop=Unit_Title Drop=Unit_Attempt_Status);
        Set Projdata;
        If cmiss(of _all_) Then Delete;
        If Final_Mark > 100 Then Delete;
        If Final_Grade = 'FAB' Then Delete;
        If Final_Grade = 'FNQ' Then Delete;
Run;

Proc Sort Data = PData;
        By Index_Key;
Run;

/* Create empty tables */
Data AllYears;
        Set PData;
        If index(Unit_Code, 'APPM') Then Delete;
        If index(Unit_Code, 'MATH') Then Delete;
Run;

Data AppmOnly;
        Set PData;
        If index(Unit_Code, 'APPM') Then Delete;
        If index(Unit_Code, 'MATH') Then Delete;
Run;

Data MathAppm;
        Set PData;
        If index(Unit_Code, 'APPM') Then Delete;
        If index(Unit_Code, 'MATH') Then Delete;
Run;
            }
```

20

```
Data AllStudents;
      Set PData;
      If index(Unit_Code, 'APPM') Then Delete;
      If index(Unit_Code, 'MATH') Then Delete;
Run;

Data AppmOnlyGrouped (Drop= Calendar_Instance_Year Drop= Year_Of_Study Drop= Unit_Code);
      Set PData;
      If index(Unit_Code, 'APPM') Then Delete;
      If index(Unit_Code, 'MATH') Then Delete;
Run;

Data MathAppmGrouped (Drop= Calendar_Instance_Year Drop= Year_Of_Study Drop= Unit_Code);
      Set PData;
      If index(Unit_Code, 'APPM') Then Delete;
      If index(Unit_Code, 'MATH') Then Delete;
Run;

Data AppmOnlyPass (Drop= Calendar_Instance_Year Drop= Year_Of_Study Drop= Unit_Code);
      Set PData;
      If index(Unit_Code, 'APPM') Then Delete;
      If index(Unit_Code, 'MATH') Then Delete;
Run;

Data AppmOnlyRepeat (Drop= Calendar_Instance_Year Drop= Year_Of_Study Drop= Unit_Code);
      Set PData;
      If index(Unit_Code, 'APPM') Then Delete;
      If index(Unit_Code, 'MATH') Then Delete;
Run;

Data MathAppmPass (Drop= Calendar_Instance_Year Drop= Year_Of_Study Drop= Unit_Code);
      Set PData;
      If index(Unit_Code, 'APPM') Then Delete;
      If index(Unit_Code, 'MATH') Then Delete;
Run;

Data MathAppmRepeat (Drop= Calendar_Instance_Year Drop= Year_Of_Study Drop= Unit_Code);
      Set PData;
      If index(Unit_Code, 'APPM') Then Delete;
      If index(Unit_Code, 'MATH') Then Delete;
Run;


/* Perform data transformations and table creation */
Proc SQL;

        Insert into AllYears
        Select * from PData as p1
        Where p1.Index_Key in (
                Select p2.Index_Key from PData as p2
                Where p2.Year_Of_Study = 'YOS 1'
                )
        and p1.Index_Key in (
                Select p3.Index_Key from PData as p3
                Where p3.Year_Of_Study = 'YOS 2'
                )
        and p1.Index_Key in (
                Select p4.Index_Key from PData as p4
                Where p4.Year_Of_Study = 'YOS 3'
                );

        Insert into MathAppm
        Select * From AllYears as a1
        Where a1.Index_Key in (
                Select a2.Index_Key from AllYears as a2
                Where a2.Unit_Code like 'MATH3%'
                )
```

21

```
        and a1.Index_Key in (
            Select a3.Index_Key from AllYears as a3
            Where a3.Unit_Code like 'APPM3%'
            );

    Insert into AppmOnly
    Select * From AllYears as a1
    Where a1.Index_Key not in (
            Select a2.Index_Key from AllYears as a2
            Where a2.Unit_Code like 'MATH3%'
            )
        and a1.Index_Key in (
            Select a3.Index_Key from AllYears as a3
            Where a3.Unit_Code like 'APPM3%'
            );

    Insert into AllStudents
    Select * From MathAppm
    Union
    Select * From AppmOnly;

/* Group students by their index keys */
    Insert into MathAppmGrouped
    Select Distinct Area, Index_Key, Age, Sex, Avg(Final_Mark),
    'Pass' From MathAppm as a1
    Where a1.Index_Key not in (
            Select a2.Index_Key from MathAppm as a2
            Where a2.Final_Grade like 'F%'
            )
        Group By a1.Index_Key;

    Insert into MathAppmGrouped
    Select Distinct Area, Index_Key, Age, Sex, Avg(Final_Mark),
    'Repeat' From MathAppm as a1
    Where a1.Index_Key in (
            Select a2.Index_Key from MathAppm as a2
            Where a2.Final_Grade like 'F%'
            )
        Group By a1.Index_Key;

    Insert into AppmOnlyGrouped
    Select Distinct Area, Index_Key, Age, Sex, Avg(Final_Mark),
    'Pass' From AppmOnly as a1
    Where a1.Index_Key not in (
            Select a2.Index_Key from AppmOnly as a2
            Where a2.Final_Grade like 'F%'
            )
        Group By a1.Index_Key;

    Insert into AppmOnlyGrouped
    Select Distinct Area, Index_Key, Age, Sex, Avg(Final_Mark),
    'Repeat' From AppmOnly as a1
    Where a1.Index_Key in (
            Select a2.Index_Key from AppmOnly as a2
            Where a2.Final_Grade like 'F%'
            )
        Group By a1.Index_Key;

Run;

Proc Sort Data = MathAppmGrouped;
        By Index_Key;
Run;

Proc Sort Data = AppmOnlyGrouped;
        By Index_Key;
Run;
```

```
Proc SQL;

/*Create and insert into summary table where students are grouped by their index keys,
marked as 'Pass' or 'Repeat' and as 'MathAppm' or 'AppmOnly' */

        Create Table SummaryTbl (
                Index_Key int,
                Area varchar,
                Age varchar,
                Sex varchar,
                MajorType varchar,
                FinalMark num,
                Result varchar
                );


        Insert into MathAppmPass
        Select * From MathAppmGrouped as a1
        Where a1.Index_Key in (
                Select a2.Index_Key from MathAppmGrouped as a2
                Where a2.Final_Grade like 'P%'
                );

        Insert into MathAppmRepeat
        Select * From MathAppmGrouped as a1
        Where a1.Index_Key in (
                Select a2.Index_Key from MathAppmGrouped as a2
                Where a2.Final_Grade like 'R%'
                );

        Insert into AppmOnlyPass
        Select * From AppmOnlyGrouped as a1
        Where a1.Index_Key in (
                Select a2.Index_Key from AppmOnlyGrouped as a2
                Where a2.Final_Grade like 'P%'
                );

        Insert into AppmOnlyRepeat
        Select * From AppmOnlyGrouped as a1
        Where a1.Index_Key in (
                Select a2.Index_Key from AppmOnlyGrouped as a2
                Where a2.Final_Grade like 'R%'
                );

        Insert Into SummaryTbl
        Select Index_Key, Area, Age, Sex, 'MathAppm', Final_Mark, 'Pass' From MathAppmPass;

        Insert Into SummaryTbl
        Select Index_Key, Area, Age, Sex, 'MathAppm', Final_Mark, 'Repeat' From MathAppmRepeat;

        Insert Into SummaryTbl
        Select Index_Key, Area, Age, Sex, 'AppmOnly', Final_Mark, 'Pass' From AppmOnlyPass;

        Insert Into SummaryTbl
        Select Index_Key, Area, Age, Sex, 'AppmOnly', Final_Mark, 'Repeat' From AppmOnlyRepeat;


Run;

Proc Sort Data=SummaryTbl;
        By Index_Key;
Run;

/*Create more blank tables */
Data ScienceFac;
        Set SummaryTbl;
        If index(Result, 'Pass') Then Delete;
```

23

```
            If index(Result, 'Repeat') Then Delete;
Run;

Data CommerceFac;
        Set SummaryTbl;
        If index(Result, 'Pass') Then Delete;
        If index(Result, 'Repeat') Then Delete;
Run;

Data EngineeringFac;
        Set SummaryTbl;
        If index(Result, 'Pass') Then Delete;
        If index(Result, 'Repeat') Then Delete;
Run;

Data MaleStudents;
        Set SummaryTbl;
        If index(Result, 'Pass') Then Delete;
        If index(Result, 'Repeat') Then Delete;
Run;

Data FemaleStudents;
        Set SummaryTbl;
        If index(Result, 'Pass') Then Delete;
        If index(Result, 'Repeat') Then Delete;
Run;

Data AgeUnder25;
        Set SummaryTbl;
        If index(Result, 'Pass') Then Delete;
        If index(Result, 'Repeat') Then Delete;
Run;

Data Age25Up;
        Set SummaryTbl;
        If index(Result, 'Pass') Then Delete;
        If index(Result, 'Repeat') Then Delete;
Run;

/* Group students according to field , sex and age */
Proc SQL;

        Insert into ScienceFac
        Select * From SummaryTbl as a1
        Where a1.Index_Key in (
                Select a2.Index_Key from SummaryTbl as a2
                Where a2.Area = 'Science'
                );

        Insert into CommerceFac
        Select * From SummaryTbl as a1
        Where a1.Index_Key in (
                Select a2.Index_Key from SummaryTbl as a2
                Where a2.Area = 'Commerce'
                );

        Insert into EngineeringFac
        Select * From SummaryTbl as a1
        Where a1.Index_Key in (
                Select a2.Index_Key from SummaryTbl as a2
                Where a2.Area = 'Engineer'
                );

        Insert into MaleStudents
        Select * From SummaryTbl as a1
        Where a1.Index_Key in (
                Select a2.Index_Key from SummaryTbl as a2
                Where a2.Sex = 'M'
```

24

```
        );

Insert into FemaleStudents
Select * From SummaryTbl as a1
Where a1.Index_Key in (
        Select a2.Index_Key from SummaryTbl as a2
        Where a2.Sex = 'F'
        );

Insert into AgeUnder25
Select * From SummaryTbl as a1
Where a1.Index_Key in (
        Select a2.Index_Key from SummaryTbl as a2
        Where a2.Age like '1%' or a2.Age = '20' or a2.Age = '21' or a2.Age = '22'
        or a2.Age = '23' or a2.Age = '24'
        );

Insert into Age25Up
Select * From SummaryTbl as a1
Where a1.Index_Key in (
        Select a2.Index_Key from SummaryTbl as a2
        Where a2.Age = '25' or a2.Age = '26' or a2.Age = '27' or a2.Age = '28'
        or a2.Age = '29' or a2.Age like '3%' or a2.Age like '4%' or a2.Age like '5%'
        );

Create Table SummaryTblAgeTests (
        Index_Key int,
        Area varchar,
        Age varchar,
        Sex varchar,
        MajorType varchar,
        FinalMark num,
        Result varchar
        );

Insert Into SummaryTblAgeTests
Select  s1.Index_Key, s1.Area, 'Under25', s1.Sex, s1.MajorType,
s1.FinalMark, s1.Result From SummaryTbl as s1
Where s1.Index_Key in (
        Select s2.Index_Key From AgeUnder25 as s2
        );

Insert Into SummaryTblAgeTests
Select  s1.Index_Key, s1.Area, '25Up', s1.Sex, s1.MajorType,
s1.FinalMark, s1.Result From SummaryTbl as s1
Where s1.Index_Key in (
        Select s2.Index_Key From Age25Up as s2
        );

Run;

Proc Sort Data = SummaryTblAgeTests;
        By Index_Key;
Run;

ods noproctitle;
ods graphics / imagemap=on;


/* produce frequency distribution */
proc univariate data=WORK.ALLSTUDENTS;
        ods select Histogram;
        var Final_Mark;
        histogram Final_Mark;
run;

/* Calculate descriptive statistics */
```

25

```
Proc Means Data = AllStudents Mean Stddev Var Q1 Q3 Median Mode Max Min Range Skewness Kurtosis;
          Var Final_Mark;
          Title Overall;
Run;

Proc Means Data = MathAppm Mean Stddev Var Q1 Q3 Median Mode Max Min Range Skewness Kurtosis;
          Var Final_Mark;
          Title MathAppm;
Run;

Proc Means Data = AppmOnly Mean Stddev Var Q1 Q3 Median Mode Max Min Range Skewness Kurtosis;
          Var Final_Mark;
          Title AppmOnly;
Run;

Proc Means Data = MathAppmPass Mean Stddev Var Q1 Q3 Median Mode Max Min Range Skewness Kurtosis;
          Var Final_Mark;
          Title MathAppmPass;
Run;

Proc Means Data = MathAppmRepeat Mean Stddev Var Q1 Q3 Median Mode Max Min Range Skewness Kurtosis;
          Var Final_Mark;
          Title MathAppmRepeat;
Run;

Proc Means Data = AppmOnlyPass Mean Stddev Var Q1 Q3 Median Mode Max Min Range Skewness Kurtosis;
          Var Final_Mark;
          Title AppmOnlyPass;
Run;

Proc Means Data = AppmOnlyRepeat Mean Stddev Var Q1 Q3 Median Mode Max Min Range Skewness Kurtosis;
          Var Final_Mark;
          Title AppmOnlyRepeat;
Run;

Proc Means Data = ScienceFac Mean Stddev Var Q1 Q3 Median Mode Max Min Range Skewness Kurtosis;
          Var FinalMark;
          Title ScienceFac;
Run;

Proc Means Data = CommerceFac Mean Stddev Var Q1 Q3 Median Mode Max Min Range Skewness Kurtosis;
          Var FinalMark;
          Title CommerceFac;
Run;

Proc Means Data = EngineeringFac Mean Stddev Var Q1 Q3 Median Mode Max Min Range Skewness Kurtosis;
          Var FinalMark;
          Title EngineeringFac;
Run;

Proc Means Data = MaleStudents Mean Stddev Var Q1 Q3 Median Mode Max Min Range Skewness Kurtosis;
          Var FinalMark;
          Title MaleStudents;
Run;

Proc Means Data = FemaleStudents Mean Stddev Var Q1 Q3 Median Mode Max Min Range Skewness Kurtosis;
          Var FinalMark;
          Title FemaleStudents;
Run;

Proc Means Data = AgeUnder25 Mean Stddev Var Q1 Q3 Median Mode Max Min Range Skewness Kurtosis;
          Var FinalMark;
          Title AgeUnder25;
Run;

Proc Means Data = Age25Up Mean Stddev Var Q1 Q3 Median Mode Max Min Range Skewness Kurtosis;
          Var FinalMark;
          Title Age25Up;
```

```
Run;


/* Perform chi-squared tests */
Proc Freq Data = SummaryTbl;
        Tables MajorType * Result / chisq;
        Title 'Chi-squared test for MathAppm vs AppmOnly';
Run;

Proc Freq Data = SummaryTbl;
        Tables Sex * Result / chisq;
        Title 'Chi-squared test for Male vs Female';
Run;

Proc Freq Data = SummaryTbl;
        Tables Area * Result / chisq;
        Title 'Chi-squared test for Different Fields';
Run;

Proc Freq Data = SummaryTblAgeTests;
        Tables Age * Result / chisq;
        Title 'Chi-squared test for Different Age';
Run;


/* Create sample data from population for two-tailed t-tests */
Proc SQL;
        Create Table TwoPops
        (Major char(5),
        Final_Mark num);

        Insert into TwoPops
        Select 'A', Final_Mark from AppmOnly
        Where Index_Key < 7202;

        Insert into TwoPops
        Select 'MA', Final_Mark from MathAppm
        Where Index_Key < 6246;

        Create Table FacTest
        (Area char(5),
        Final_Mark num);

        Insert into FacTest
        Select 'S', FinalMark from ScienceFac
        Where Index_Key < 8232;

        Insert into FacTest
        Select 'O', FinalMark from CommerceFac
        Where Index_Key < 15064;

        Insert into FacTest
        Select 'O', FinalMark from EngineeringFac
        Where Index_Key < 10308;

        Create Table SexTest
        (Sex char(5),
        Final_Mark num);

        Insert into SexTest
        Select 'M', FinalMark from MaleStudents
        Where Index_Key < 7313;

        Insert into SexTest
        Select 'F', FinalMark from FemaleStudents
        Where Index_Key < 9838;
```

27

```
        Create Table AgeTest
        (Age char(5),
        Final_Mark num);

        Insert into AgeTest
        Select 'U', FinalMark from AgeUnder25
        Where Index_Key < 9472;

        Insert into AgeTest
        Select 'O', FinalMark from Age25Up
        Where Index_Key < 11389;

Run;

/* Perform two-tailed t-tests */
Proc TTest Data=TwoPops Sides=2 Alpha=0.05 h0=0;
        Title "Two sample t-test for student marks";
        Class Major;
        Var Final_Mark;
Run;

Proc TTest Data=FacTest Sides=2 Alpha=0.05 h0=0;
        Title "Two sample t-test for student marks and faculties";
        Class Area;
        Var Final_Mark;
Run;

Proc TTest Data=SexTest Sides=2 Alpha=0.05 h0=0;
        Title "Two sample t-test for student marks and genders";
        Class Sex;
        Var Final_Mark;
Run;

Proc TTest Data=AgeTest Sides=2 Alpha=0.05 h0=0;
        Title "Two sample t-test for student marks and ages";
        Class Age;
        Var Final_Mark;
Run;


/* Proc Corr Data = SummaryTbl fisher; */
/*      Var Age FinalMark; */
/* Run; */
```